Promoting the Essence of WASHOKU with Seven Vegetables

Poems & Photos by Iroha HARA

詩歌で伝える和食七つ菜
(しいか)(つた)(わしょく)(ななつな)

この国で七つ菜を栽培販売する100万の営農家を応援します
「和食七つ菜」は登録商標です

この歌に導かれて　A historical poem as prologue

800年前 日本に春夏秋冬の「国おこし歌」があった

春の朝に桜花を見よう。夏の夕に山鳥の声を聞こう。秋の夜に月の満ち欠けを見よう。冬の早朝には山の雪を見よう。日本人は自然と共生し、こんな四季の美のなかで暮らそう、と慈円さまは和歌で提唱した。800年前詠まれた見事な「国おこし歌」に導かれて私たちふたりは「四季の美を愛でる家庭の和食」「禅の教えに基づく食事」を詩歌と写真で表現しました。

A poem for nation-building in Japan 800 years ago

"Let's admire cherry blossoms in the mornings of spring. Let's listen to the voices of wild birds in the evenings of summer. Let's observe the phases of the moon in the nights of autumn. And let's view snowy mountains in the early mornings of winter." This historical poem calls on Japanese people to cohabit with nature and to enjoy living in the scenic beauty of four seasons. Guided by this historical poem, we hope to present to you "*Washoku* at home to enjoy four seasons" and "meals based on the concepts of Zen" with original poems and photos.

花

春のやよひの
あけぼのに
四方の山べを
見わたせば
花ざかりかも
白雲の
かからぬ峰こそ
なかりけれ

ほととぎす

花たちばなも
にほふなり
軒のあやめも
かをるなり
ゆふぐれざまの
五月雨に
山ほととぎす
なのりして

Wishing to communicate through poems and photos about *Washoku* that appreciates motherly mountains and water as well as the seasonal "Seven Vegetables"

Mountains more than 1000 meters high foster deciduous woodlands and cultivate mushroom vegetables as well as bean and fruit vegetables, by providing water rich in nutrients to their lower areas. They grow rice, root and stem vegetables, by wetting farm fields there. They also produce leaf and sprout vegetables, by misting their lower reaches. Mountains and the water running from them seem to create the "Seven Vegetables" effortlessly. The nutrients of these mountains nurture sweet fish and trout in the rivers, and cultivate seaweed and oysters in the seas. There are 724 motherly mountains more than 1000 meters high in Japan. With our thanksgiving to these mountains and water of Japan and its seasonal "Seven Vegetables," we hope to communicate to you *Washoku* meals enjoyed at home in the morning and evening using these vegetables through poems and photos.

四季の「七つ菜」に感謝する和食を詩歌と写真で伝えたい　母なる日本の山と水

標高1000mを超える山は、栄養豊富な水を下流に一年中供給し、落葉樹林を育て茸菜や実菜を作る。田畑を潤し米菜や根菜や茎菜を作る。里に霧を発生させ葉菜や蕾菜を作る。新鮮な「七つ菜」を易々作る。山の栄養は川で鮎や鱒を、海で海苔や牡蠣を育てる。1000mを超える母なる山が、日本には724座もある。この国の山と水と四季の「七つ菜」に感謝し朝な夕な家庭でいただく和食の膳。それを伝えたい。

月

秋の初めに
なりぬれば
今年も半ばは
すぎにけり
わがよふけ行く
月影の
傾ぶく見るこそ
あはれなれ

雪

冬の夜さむの
朝ぼらけ
ちぎりし山路に
雪ふかし
つかねども
心の跡は
おもひやるこそ
あはれなれ

慈円大僧正の国おこし歌

鳥も樹も
花も菜も
親も子も
生きものは皆
等しく菩薩さま
此れ
この国の
生活上手の
極意かな

Birds and trees,
flowers and vegetables,
or parents and children,
all living things are
equally Bodhisattvas.
This belief could be
a key to
living well
in Japan.

七つ菜・茶の若葉

ありがとう
ありがとう
在るがままに
ながれるように
　此れ
　この国の
　諸道上手の
　極意なり

Let's accept,
appreciate,
and admire
living things
as they are naturally.
This attitude should be
a secret for
good Japanese ways
of living.

伊路次

随筆家 松本 章男

『慈円さまの春夏秋冬 国おこしの歌に導かれて夫婦で膳を作り写真を撮り詩を書き愉快です』と言う原氏は、お家に伝わる器で膳をしつらえわが子を愛しむ如く語りかけ「接写で撮影」する。こんな創造的私生活が出来ることは驚きです。日本人の優しさ・生き方を世界に伝える本です。

版画家 翁倩玉 ジュディ・オング

私は日本の花や建築を版画にするのが好きで日本の食材で体に良い料理を作るのも好きです。原伊路波氏の「詩歌で伝える和食七つ菜」本は外国の方々も知りたがる日本の家庭の食膳です。四季折々の国産伝統野菜の美しさと、私たちの体にやさしい七つ菜のチカラが伝わってきます。

春の膳
- 国おこし歌 2
- この国の極意 4
- 推薦の辞 6
- 花わさび 8
- 本わさび 14
- 花どんこ 20
- 原木椎茸 26
- 一寸空豆 32
- 春みかん 38

夏の膳
- 紅茶の漆器 44
- 1番摘み茶 52
- 碾茶と抹茶 58

ふたり膳
- 2茶2食 66
- 台所で禅 68
- 国産七つ菜 70
- ふるさと大根 74

学校法人 滋慶学園グループ
総長 浮舟 邦彦

美しい野菜の姿をこわす加工や調理をしないで
『在るがままにながれるように 此れこの国の
諸道上手の極意なり』と詠んだ原伊路波氏は
クリエーティブデザイナーと教育アドバイザー
「実学の道」「奉仕の道」を四十年歩み続けました。
作為技術に走らず自然体の著作は原氏らしい。

ふたり膳
　愛しき野菜　76
　水で料理　78
　天然の三味　80
　先祖の器　82

秋の膳
　千年小菊　84
　みのり穂　92
　ふうふ福豆　100
　黒豆舟遊び　106
　本くずの絵　112

冬の膳
　霧の小蕪　120
　里芋の花筏　126
　万能ゆず酢　130
　立花ねぎ　134
　お米おせち　140
　蕾菜かゆ　148
　四弁の黄花　152

あとがき
　花を描く　156
　この国に感謝　159

春の膳① 花わさび　Spring ① Hana-wasabi, flowering delicate-stem wasabi

清らかな山の水で育つ　日本の代表野菜

私たちふたりは 四季を愛でる家庭WASHOKUの第一番目に早春のわさびを紹介します。「わさび」は学名Wasabi Japonica。1000年以上伝わる日本原産の野菜。清らかな山の湧き水で育ち、春いちばん早く白い花を咲かせます。若い蕾菜を「花わさび」と呼び、本体下部の茎菜を「本わさび」と呼ぶ。野菜類を余さずに全部いただくのがWASHOKUの真髄。美味しいアブラナ科「本わさび」は、日本の代表野菜。

Representative Japanese vegetable grown with pure mountain water

We would like to introduce to you *Hana-wasabi* or the flowering delicate stem-part of wasabi as a leading vegetable for *Washoku* at home to enjoy the four seasons. The academic name of wasabi is *Wasabi Japonica*, a vegetable that originated from Japan 1000 years ago. It is grown with pure spring water from mountains, and its tiny white flowers bloom in the earliest spring. The delicate stem of wasabi with white flowers is called *Hana-wasabi* (*Hana* means "flower" in Japanese), while the main stem is named *Hon-wasabi* (*Hon* means "authentic" in Japanese). It is the essence of *Washoku* to enjoy the whole of a vegetable without any waste. Delicious *Hon-wasabi*, a brassica vegetable, has always been one of the vegetables that represent Japan.

わさび田（だ）の
真白（ましろ）き花摘（はなつ）み
ああ楽（たの）し
雪解（ゆきど）けの水（みず）
ああ涼（すず）し

春の膳①花わさび

春に生まれし
　花わさび
秋さく菊花に
　迎えられ
こころひらく
真白き花ひらく
　しつらえ楽し
美の出会い

春秋(はるあき)の出会い

雪解け湧水で育った「花わさび」お供え花　光琳菊花図の枕屏風と 金銀蒔絵黒漆塗お重の花器

春の膳①花わさび

咲くざくざく

育ての親は
1000m余の
山から湧く水
花わさび
噛む音はずむ
心はずむ
1000年余の
香りつーん

菜花のように霧蒸した花わさびを、蕎麦にたっぷりと乗せて「花わさび添え・栃の実そば」

春の膳①花わさび

摺った本わさびに 香りあり 薬効あり

年輪が重なる下部の緑色茎菜がWasabi Japonica「本わさび」です。この国の野菜は金気を嫌うので私たちは写真のように「本わさび」をサメ皮かセラミックでゆっくり摺り下ろします。野菜の香りと薬効を引き出すのがWASHOKUの真髄。摺り下ろした本わさびは、米・野菜・魚・肉の万能香辛料で抗菌・消化・血栓予防など薬効が認められています。

香(かお)りとどけば
心(こころ)がふわり
本(ほん)わさび
からだ温(ぬ)くぬく
わさび酒(ざけ)

Grated authentic wasabi rich in flavor and medicinal effects

The bottom part of this green stem vegetable with ring-like traits is *Hon-wasabi* or authentic main-stem wasabi. As vegetables grown in Japan are not suited to the use of metallic utensils, we carefully grate *Hon-wasabi* with a dried shark skin or a ceramic utensil. It is the essence of *Washoku* to bring out the flavor and medicinal effects of vegetables to the fullest extent. Grated *Hon-wasabi* is a versatile spice for rice, other vegetables, fish and meat, and is generally acknowledged to have various medicinal effects such as anti-germ, digestive and blood-clot preventive.

春の膳 ② 本わさび　Spring ② Hon-wasabi, authentic main-stem wasabi

春の膳②本わさび

わが家の台所は春のエネルギーで満ちる　ふるさと伊豆から届いた「本わさび」と、右頁上は兄の手作り「わさび漬」

わさびに作法あり

お茶に作法あり
丁寧に石うす
挽きて香りあり
薬効あり

湧水の本わさび
丁寧にサメ皮
摺りて香りあり
薬効あり

サメ皮で摺った本わさび匙1杯を 温めた純米酒140ccに溶いた薬効酒「春わさび酒」

左上は 本わさびの葉と茎で作る「春わさび寿司」　手前は 純米酒と本わさびで作る「春わさび酒」1人 70cc

本わさびかゆ

道元禅師さまは
　徳を説く
お粥に作法あり
　薬効あり
山の水神さまも
　徳を説く
わさびに香あり
　薬効あり

かゆ十徳（血色・力・寿命・楽・言葉・清腸・風邪・満足・渇消・便通）の効果に
わさび六徳（消化・食欲増進・抗菌・下痢止め・血栓予防・美容）の効果も…
伊豆や安曇野の春を届けてくれる「本わさびかゆ」はふたりの薬効膳です

ふたば葵の金銀蒔絵黒漆塗半月盆に わさび葉を浮かべ「しらす 本わさび」と「つぶ味噌 本わさび」で粥をいただく

春の膳②本わさび

春の膳 ③ 花どんこ Spring ③ Hana-donko, flowery Shiitake mushrooms

梅から桜咲くころが美味しい 春の宝石

「花どんこ椎茸」はクヌギ・ナラ・シイなど日本の自然林で2年かけて育てる茸菜です。冬から早春に採取し乾燥させ、小粒で表面が花様の茸。ふるさと伊豆市が産地で、宝石のような花どんこ椎茸が届くと我が家に春が来ます。梅から桜の咲く頃が香りよく小粒なのに「戻し汁」が素晴らしい。山と海の食材を合わせて「旨みUMAMI」を倍加させる、がWASHOKUの真髄。ぜひ「花どんこ戻し汁」を。

Gem-like spring mushrooms to be most delicious in the seasons of plum to cherry blossoms

Flowery Shiitake mushrooms are grown over two years in natural groves such as sawtooth oak. They are small mushrooms with flowery patterns on top, and are usually picked and dried from winter to early spring. We feel that spring arrives home when these gem-like shiitake mushrooms are delivered from *Izu, Shizuoka* where they are produced. Their flavor becomes the best in the seasons of plum to cherry blossoms, making their soaking liquid quite mouth-watering. It is the essence of *Washoku* to enhance *Umami* flavor by combining ingredients from mountains and seas. We hope that you will also enjoy these flowery shiitake mushrooms cooked with their soaking water at your home.

ふたりして
冬(ふゆ)にあつめた
花(はな)どんこ
桜花(おうか)に見紛(みまご)う
早春(そうしゅん)の蝶(ちょう)

春の膳③花どんこ

四季の「花鳥」を愛し、里山の「野菜」に感謝する…WASHOKUは「山と水と七つ菜の国」に伝わる生活文化です

春の宝石

梅から
さくらの
間にとどく

ふるさとの
香りのせて
花どんこ

春の膳③花どんこ

花どんこ戻し汁

寒流そだち
真こんぶ
寒山うまれ
原木しいたけ
ふたつの旨み
ひと口で
これ口福
春のふたり膳

原木椎茸「花どんこ」4個を水に浸した「戻し汁」に 天然真こんぶ1片で「最良の出し」に

春の膳③花どんこ

よもぎ生麩焼と芽ねぎを添えて「花どんこ戻し汁」 平椀は 青漆塗金銀色蒔絵 絵替花図

1000年前から日本を代表する旨み食材

道元さまの時代から、原木椎茸は日本の輸出食材。その姿と旨みは「山の鮑」と呼ばれる。私たちの先祖は自然の山菜や果実や花を1000年かけて大きく美しく育ててきました。同時に食材として旨みが増すように味も改良。大切に育てた野菜だから「姿を残す」「旨みを残す」ひと口大でいただくのがWASHOKUの真髄。早春は生椎茸「山のあわび」、桜花が過ぎたら干し椎茸「戻しあわび」で、どうぞ。

クヌギの森に
ふたりして
山（やま）の鮑（あわび）を
むかえゆく
春（はる）のよろこび

Japanese representative food ingredient for *Umami* since 1000 years ago

Shiitake mushrooms grown on natural logs have been an ingredient for export for 1000 years in Japan. Their appearance and tastiness are outstanding enough to be comparable with those of a beard-clam. Our ancestors have cultivated wild vegetables, fruits and flowers for size and beauty over 1000 years, and also improved their tastes as ingredients to intensify *Umami*-flavors. It is the essence of *Washoku* to enjoy bite-sized, carefully-grown vegetables which maintain their original shape, texture and taste. Please enjoy fresh shiitake mushrooms in early spring and dried ones after late spring.

春の膳④ 原木椎茸 Spring ④ Shiitake mushrooms grown on natural logs

日本の原風景自然林はクヌギ・ナラ・シイ・クリが主役　「原木椎茸」はこれら広葉樹の枯れ木に発芽して育つ

春の膳④原木椎茸

山のあわび

春のやよい
明けそめし
クヌギの山べを
見わたせば
今こそ摘み頃
食しごろ
森の香りを
蒸していただく

山で摘んだ原木生椎茸は直径 8.5cm × 肉厚 3.5 cm　瑞々しい森の香りを逃さないよう 切らずに蒸す

あわび形の青織部向付に蒸した原木椎茸を盛って愉しむ「山のあわび春野菜蒸し」 青織部香合に「つぶ味噌」添えて

春の膳④原木椎茸

戻しあわび

山の鮑
すこし乾くと
滋養高まり
旨み増す

原木干し椎茸
水で戻した
その身の厚み
汁の深み

原木干し椎茸を水に戻すと直径8cm×肉厚3cm ぷりぷりの「食感」も「戻し汁」も海の鮑より上品

原木干し椎茸で春らしく「山のあわび戻し汁煮・菜花つぼみあんかけ」 小鉢と深向付は 鳴海織部白花図

春の膳④原木椎茸

日本の少年は　平たい薄緑の豆が好き

春の終りから初夏に、山の樹は産毛がはえた薄緑色の葉が芽吹き、里の畑は薄緑色の大粒豆が実ります。英語で「幅広い豆 broad beans」と表記するように平たく大きい。幅が３cm余あるので、日本では「一寸空豆」と呼ばれる。美味しい期間は完熟前の３日間です。旬野菜の美や味を、水でひき出すのがWASHOKUの真髄。私は空豆を、蒸していただく。

ふわっふわの
ふとんに抱（いだ）かれ
おおきく育（そだ）つ
春（はる）の空豆（そらまめ）
一寸（いっすん）ぼうし

春の膳⑤　一寸空豆　Spring ⑤ One-inch broad beans

Flat olive-green beans favored by Japanese boys

From the end of spring to the beginning of summer, hairy light-green leaves appear on trees in mountains, while large olive-green beans grow in fields at the bottom of these mountains. Broad beans are large flat beans of three centimeters in width, called "One-inch broad beans" in Japan. They become their most luscious for three days just before their full ripeness. It is the essence of *Washoku* to boil or steam early-picked seasonal vegetables only with water. We enjoy broad beans by steaming them.

春の膳⑤ 一寸空豆

33

空豆は
　みな
空に向かって
　花咲き
　おおきく
　　育つ

少年は
　みな
空豆を
ポケットに
　大志を
　　いだく

春の膳⑤　一寸空豆

一寸空豆すり流し

半世紀がすぎ
初老の少年
こよいも　空豆
いただく

おおつぶ旨し
汁はもっと
　　　うれし
こころ温まる
花冷えの春

一寸の愛称どおり3cm余りの大粒　ふわっふわの布団に抱かれたまま土鍋に並べ霧蒸しに

春いろ美しい「一寸空豆すり流し椀 旨み出し汁豆乳仕立て」 椀は 赤漆塗金彩蒔絵 松笠図

春の膳⑤ 一寸空豆

5月に摘む 白い花 オレンジ色の果実

日本の春果実は「紅甘夏 Beni-ama-natsu」みかん。初夏に出荷するから夏ダイダイとも呼ばれるが旬は春の後半です。春に摘んで 春に食するから私は「春みかん」と呼ぶ。甘いのに すがすがしい。今年の果実が摘み頃のとき、来年の花がいっぱい咲く。香りまで楽しむのがWASHOKUの真髄。春みかん=紅甘夏の白い花からつくる「香り水」はこころが 静まり、枕に振ると ぐっすり眠れます。

Orange fruit with white flowers to be picked in May

One spring fruit of Japan is the sweet ruby summer orange, a variety of Watson Pomelo. Although the orange is usually delivered to the market in the beginning of summer as its name suggests, it comes into season in the latter half of spring. Therefore, we prefer to call it the "Spring Orange," picking and tasting it in its best season. It is sweet but refreshing. When the orange reaches its best season to be picked in spring, its mother tree blooms plenty of white flowers for bearing fruits the next spring. It is the essence of *Washoku* to enjoy not only the taste but also the scent of ingredients. Aromatic water made from pesticide-free "Spring Orange" gives us peace of mind when sprayed into the air, as well as calm sleep when sprayed onto pillows.

かおる　香(かお)居る
五弁(ごひら)の花(はな)から
春(はる)の蜜柑(みかん)の
香(かお)り水(みず)

春の膳⑥ 春みかん Spring ⑥ Spring Orange

今年の果実がオレンジ色して摘み頃のとき 同じ春蜜柑の樹に 来年果実になる白い五弁花がいっぱい咲く 香り極上

春の膳⑥春みかん

[1] 香りを放つ
五弁の花を摘む

[2] ビーカーに
花と水を入れる

40

La première 香り水

日本の冬に耐え
　香り蓄え
　甘み蓄え

春みかん花と実
披露の宴に
フランス婦人
　花は炊かれ
　蒸気でのぼり
　水滴で落ち

みかんの香り水
顔よせ吸って
仏語ひとつ

ラ・プルミエール
貴女いちばん
（すばらしい）

[4] 2時間で 17cc の香り水

[3] フラスコを置き 火をつけ 香りの水蒸気を集める

Le premier 春みかん

日本を愛する
老婦人に敬意
黄金フルーツ膳
寒い冬甘み蓄え
　春になる
　果実ひとつ
堂々たる果肉
　手にのせ
　仏語ひとつ
ル・プルミエ
貴方いちばん
（たくましい）

零下の厳しい冬を越し、春に花咲き果実になる　伊豆・伊予・有田・熊本など「国産春みかん」が一番

紅甘夏と三宝柑の「黄金フルーツ膳」 盛り皿は45cmマイセンアンティーク金彩磁器

春の膳⑥春みかん

日本人は「お茶時間」を大切にしてきた

WASHOKUは1日2回のお茶時間と3種の茶葉が不可欠。私たちふたりは祖父母が好んだ漆器紅茶椀で甘く色鮮やかな国産紅茶を愉しみます。先人から伝わる家庭食器は、保温性芸術性に優れた木製漆器が主役。何度も塗り重ねた色漆の上に金銀彩で春秋の草花が描かれる。茶棚・茶箱・水指・茶入・お盆・菓子皿など茶道具。お膳・お重・鉢・汁椀など食事道具。いずれもjapanと冠される日本の美術工芸品です。

The "Teatime" long valued by Japanese

It is the indispensable pleasure of *Washoku* to enjoy teatimes twice a day with three kinds of tea leaves. We enjoy sweet and bright-colored tea domestically produced with the lacquer teaware favored by our grandparents. Notable among various kinds of tableware passed down from our ancestors is wooden lacquerware that is both functional in retaining moisture and aesthetically pleasing. It is made of wood with multi-layered and colorful lacquer paintings as well as with gold and silver patterns of plants and flowers. Tea utensils such as tea shelves, tea boxes, water jars, tea containers, trays and plates for sweets, as well as tableware such as platters, stacking boxes, bowls and soup cups: all are Japanese fine arts and crafts proudly called "japan."

夏の膳① 紅茶の漆器 Summer ① Lacquer tea ware

この国の美に心して
2茶（にちゃ）
3茶（さんちゃ）

抹茶も煎茶も紅茶も 新茶は良い水がいのちです 「1番摘み紅茶」を活かす水指は 黒漆塗金銀蒔絵 秋の総花図

夏の膳① 紅茶の漆器

かがやく紅茶

金で手厚く
　茶葉守る
この国の茶葉に
　チカラあり
新茶が魅せる
　金色の輪
この国の茶葉に
　チカラあり

金の茶箱で守られた1番摘み紅茶をいれると、甘く旨し　水色は澄んで「黄金の輪」かがやく

「1番摘み紅茶」を迎えた黒漆塗金蒔絵 梅菱桐葉図茶箱　内側が本金厚塗仕上げで、無農薬新茶を金のチカラで守る

夏の膳①紅茶の漆器

かるーく紅茶

香りたのし
はな絵たのし
茶椀もちあげ
何十かい

手にかるい
よき茶椀
保温も良かれ
この国の紅茶椀

英国製ボーンチャイナ（骨灰乳白色1881年）のカップ本体の軽さ66g

漆工芸の国 japan 製紅茶椀はカップ本体の軽さ48ｇ　黒漆塗金蒔絵 枝垂れ桜図お揃い絵の長角盆と紅茶椀

夏の膳①紅茶の漆器

49

ふたり紅茶

国産紅茶
ありがとう
祖父母好んだ
ティセット

ひとつの膳に
茶菓かざり
漆黒あか金
べに色くらべ

赤金
あかきん

紅茶の
べにいろ

漆黒
しっこく

本金のチカラで紅茶とゼリーの「べに色」まぶしい　菓子皿と紅茶椀は 黒漆塗金蒔絵 江戸花図

花の絵替わり漆器に 国産紅茶と紅茶ゼリー 紅茶葉自体に「ほどよい甘味」があるから、砂糖など不要です

夏の膳①紅茶の漆器

旬を蒸していただく お茶こそ和食なり

野菜を蒸すと汚れは落ち栄養は逃げない。香りは出る。温かくても冷たくても旨い。5月初旬に「1番摘み」して蒸した茶葉がWASHOKUの模範です。太陽が強く、湿度が高くなる夏は 野菜も茶も育つ。大家族の食事は大きく育ち大量に出回る旬もの＝夏摘み茶、夏採り野菜が経済的。わが家のふたり膳は旬の45日前＝小さめで柔らかい春採り野菜を、香りも味も深い春摘み茶葉を蒸していただきます。

紅茶（こうちゃ）緑茶（りょくちゃ）
ふたつ迎（むか）えて
この国（くに）の
1番摘（いちばんつ）みに
ああ夏（なつ）の香（かお）り

夏の膳② 1番摘み茶　Summer ② First-flush tea leaves

Enjoying the steamed tea leaves in season that embody *Washoku*

Steaming vegetables helps them keep their nutrients while washing away dirt, brings out flavors, and renders delicious both hot and cooled. It is the ideal of *Washoku* to enjoy the subtle flavor of "first-flush" tea leaves in early May. Both vegetables and tea leaves powerfully grow into full size during summer with its strong sunshine and high humidity. For a large family, tea leaves and vegetables that are picked in summer after growing to their fullest extent, therefore abundant in the market may be more favorable from the economic perspective. In preparing meals for two at home, we prefer to enjoy small and soft vegetables and fresh tea leaves both picked in spring, about 45 days before they come fully into season.

夏の膳② 1番摘み茶

水出し舞踏会
深蒸し茶
キミは
みどりの貴公子
甘くさわやか
日本紅茶
貴女はレディ
こよいは
水の舞踏会

わが家では茶漉し付サーバー2本使って深蒸し茶と日本紅茶の「水出し茶」を一度につくる
茶葉はお湯出しの場合より多め(15g)を入れて浄水1000cc注ぎ、冷蔵庫で一晩寝かす
朝ふたり分をサーバーからガラスポットに移し室内で温めて、2つに注いで新茶味くらべ

夏の膳②1番摘み茶

「WASHOKUの夏茶」が欲しくて 静岡や熊本の「1番摘み深蒸し茶」を水出しで試す 美しい水色が出て 甘く旨い

甘党の深蒸し茶

たばこ止め
酒ひかえ
茶菓の世界に
　迎えられ

ふたりして
豆乳ほっと茶
　初夏の
　　朝ぼらけ

夏の午後は すこし冷した「深蒸し茶あずきスイーツ」 盆は 黒漆塗金銀蒔絵 紫陽花図

夏の膳② 1番摘み茶

深蒸し茶の粉末泡立て「豆乳ほっと茶」　盆は 黒漆塗金銀蒔絵 朝顔図　棗は 白漆塗金銀蒔絵 夕顔図

- 淡萌黄 うすもえぎ
- 萌黄 もえぎ
- 常磐色 ときわいろ
- 松葉色 まつばいろ

お茶からWASHOKU全てを学べる

日本は「諸道」があり、作意技術に走らず、神聖な行為を学ぶ。伝統的WASHOKUは「七つ菜道」である。七つ菜はみな神聖な存在。生け花にして先祖に供え自然の色と味を誓う。着色せずに天然の色を引き出す。加工や調味を加えずに旨みを引き出す。わが家のふたり膳は七つ菜を先祖からの器に盛り「感謝の心と所作」でいただく。茶は「葉菜」のひとつで、とくに碾茶と抹茶からWASHOKUの心技体を学べます。

Learning everything about *Washoku* from Japanese tea leaves

Japan values many practices that teach not merely techniques but also divine acts expected to derive from the bottom of the heart. Traditional *Washoku* represents the way of enjoying the "Seven Vegetables." They are all sacred living nature. We offer these vegetables to our ancestors, pledging to safeguard their original forms, tastes and colors. We bring out their natural flavors without processing them or adding artificial colors. We eat them at home on fine Japanese tableware from our ancestors, and enjoy them with a thankful heart. Tea leaves as a leaf vegetable, especially those of *Tencha* and *Matcha*, teach the heart, technique and acts of *Washoku*.

夏の膳③ 碾茶と抹茶　Summer ③ Tencha and Matcha

淡萌黄
萌黄に常磐
碾茶の木
葉色かがやく
夏は来ぬ

赤志野茶碗に迎えられた「初夏の碾茶」の樹　右のよしず覆い下の茶園と同じ樹林のように自然に活けてお供えする

夏の膳③　碾茶と抹茶

碾(ひ)く茶と書いて

風そよそよ
木漏れ陽
ゆらゆら
覆い下の茶園
てんちゃと知る
茶の女王様
いつもお傍に
育茶マイスター

『摘み頃は新葉の表面がしっかり開いた時』と、教えられた。4月上旬から30日ほど覆いをした茶畑の新葉を、5月上旬に1年1度だけ手摘みして蒸して、揉まずに作るのが碾茶です。昔から緑茶は飲む薬草であり、碾茶の成分(アミノ酸・カテキン・テアニン・ビタミン・フッ素・食物繊維等)に薬効多く老化・糖尿・整腸・利尿・風邪・虫歯・口臭・抗菌・美容にも効果。素晴らしい素材「碾茶」を、石臼で碾いて「抹茶」が完成します。

揉まずに完成した碾茶葉

夏の膳③ 碾茶と抹茶

碾茶の茶葉(上)はとても軽い 60℃の湯を注ぐと紅茶葉のように対流する

碾茶は舞う

天空ではなつ みどり葉
ひらひらと
舞姫様のよう

湯をそそぐ
ポットのなかで
カップのなかで
てんちゃ舞う

夏の膳③ 碾茶と抹茶

63

みどり茶の女王

碾茶と抹茶
ふたつきて
夏のアトリエ
かおるなり
初夏に迎えて
はやうれし
猛暑ふたりで
のりこえる

左の「碾茶」葉を石臼で碾いて、右の「抹茶」粉が完成する　毎年初夏わが家に、ふたついっしょに届く

みどり茶の女王「抹茶」を迎える薄茶器は 黒漆塗金彩高蒔絵 竹林図中次

夏の膳③ 碾茶と抹茶

余さず 残さず 1日2食を美しく

「今日1日を無駄な時間なく生きる」と同じくらい「今日1日を無駄なく食す」ことは難しい。食材を余さずに使って1日の料理を作り、残さず2回で食べる。ふたりなら出来そうです。「今日1日を上手に働く」と同じくらい「今日1日を上手に休憩する」のは、60歳過ぎても難しい。ふたりなら出来ます。香ばしい茶を1日2回、お気に入りの器でいただく「ふたり時間」はWASHOKUがくれた贈り物です。

午後のふたり膳 七つ菜の「立花ねぎ」

朝（あさ）な余（あま）さず
夕（ゆう）な残（のこ）さず
美（び）に
こころして
2（に）茶（ちゃ）2（に）食（しょく）

ふたり膳① 2茶2食 Essence ① Two teatimes and Two meals a day

Eat beautifully only twice a day with no remains and no leftovers

It is as difficult to waste no food each day as it is to waste no time each day. We cook dishes for a day using the entire vegetables without any remains, and finish eating them in two meals without any leftovers. We can do this if we are two. It is as difficult to rest well each day as it is to work well each day, even after the age of 60. But we can do these rituals if we are two. "Time for two," enjoying fragrant tea in our favorite cups twice a day, is truly a gift from *Washoku*.

午前の茶

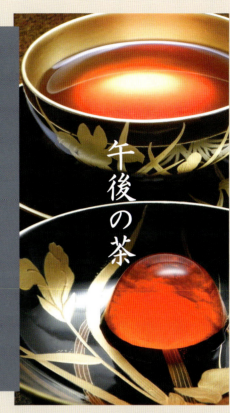

午後の茶

わが家のふたり膳は1日2茶します
この国のお茶は「和食の美の基本形」
「和食器でいただき上手」を学びます

午前の食

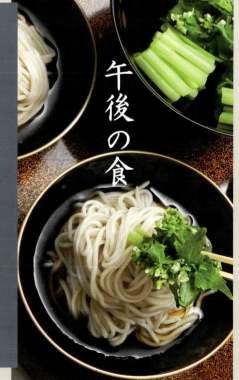

午後の食

わが家のふたり膳は1日2食します
この国の食事は「生きてる全てに感謝」
「余さず残さず」「美しい膳」を学びます

ふたり膳① 2茶2食

朝も夕も ふたりして 台所で学ぶ

「先祖から伝わる芸道を独学で学ぶ」と同じくらい「先祖から伝わる禅宗を独学で学ぶ」ことは難しい。道元禅師さまは「寺の厨房は禅修行の場。朝夕の食事作りから離れず禅を学ぼう」と説いた。わが家も朝夕ふたりで台所に立つ。いうなれば、わが家の膳はふたりの禅修行。1人が米を炊き、1人が菜を蒸し、汁を作る。食材に感謝、今日に感謝、先祖に感謝。禅修行の台所こそ、WASHOKUがくれた教室です。

菜を蒸し
米を炊き
汁を作り
厨房に居る
不離叢林の時

初夏の不離叢林は
七つ菜「新茶の出し粥」

Learning Zen in the kitchen together in both the morning and the evening

As with self-educated art transmitted from our ancestors, it is difficult for us to learn Zen they have passed down by ourselves. *Dōgen Zenji*, one of the leading Zen masters in Japan, taught that "as the kitchen of a temple offers the opportunity of practicing Zen, we keep ourselves engaged in preparing meals in the morning and the evening." We are also in the kitchen together during both the morning and the evening. Preparing meals for two is, in a sense, our way of practicing Zen. One of us cooks rice, while the other steams vegetables and makes soup. We appreciate the ingredients, today's world, and our ancestors. The kitchen for practicing Zen is the classroom offered by *Washoku*.

ふたり膳② 台所で禅　Essence ② Practicing Zen in the kitchen

わが家のふたり膳は禅の学びです
道元禅師さまは自分の台所も叢林
（修行の場）であり不離と説きます

わが家のふたり膳は台所の禅です
夫婦・姉妹・兄弟・母娘・親友お二人で
朝の叢林…夕の叢林…楽しきかな

玄米白米を炊く

夫は

二人でわさび汁を作る

妻はつぼみ菜を蒸す

毎食「七つ菜」から ひとつずつ選ぼう

人間の歯の25％は門歯で、口より大きく柔らかい五つ菜（根・茎・葉・茸・蕾）をひと口サイズに切る機能。63％は臼歯で、固い二つ菜（米・豆）を消化サイズにつぶす機能。これら88％が野菜用で、残り12％が動物の肉皮を嚙む犬歯。水が豊かな日本は、山に里に「七つ菜」が育ち、鳥獣狩猟に頼らず食べてきた幸運な国です。「七つ菜」選びがWASHOKU事始め。米料理と豆料理を決め、五つ菜を合わせるのがコツ。

bean and fruit vegetables

黒豆

黒豆（くろまめ）黒米（くろまい）
菜花（なのはな）さといも
ねぎ椎茸（しいたけ）ごぼう
この国（くに）の
七（なな）つ菜（な）に力（ちから）あり

Selecting the "Seven Vegetables" for each meal

25% of human teeth are incisors designed to cut the five large soft vegetables (roots, stems, leaves, mushrooms and sprouts) to bitesize. 63% are molars to crush the two hard vegetables (rice and beans) into a digestible texture. Therefore, 88% in total are for vegetables while only the remaining 12% are canines to chew meats. Japan, rich in water resources, is a fortunate country where the "Seven Vegetables" have abundantly grown in mountains and where the people have been able to feed themselves without relying on hunting. Selecting appropriate and favorite vegetables from the "Seven Vegetables" is a good start for *Washoku*. One tip is first to decide on dishes using rice and bean vegetables, and then to choose the other five vegetables that go well with these two dishes.

ふたり膳③ 国産七つ菜 Essence ③ Home-grown "Seven Vegetables"

ごぼう

root vegetables

菜花

sprout vegetables

黒米

rice vegetables

ねぎ

leaf vegetables

椎茸

mushroom vegetables

さといも

stem vegetables

ふたり膳③ 国産七つ菜

わが家のふたり膳は 国産七つ菜（蕾・実・米・葉・茸・茎・根）に選ばれた野菜穀物群を「偏りなく1日2回たっぷり」食べて体をつくる食事です
WASHOKUの主役は米と豆…米豆料理を決めて五菜を合わせます

つぼみな　蕾菜

- はなわさび　山葵の若い蕾菜
- なのはな　菜花
- つぼみな　蕾菜・からし菜の芽
- ふきのとう　蕗薹

sprout vegetables

みのな　実菜

- くろまめ　黒豆
- そらまめ　空豆
- ゆず　柚子
- だいず　大豆
- えだまめ　枝豆
- かぼちゃ　南瓜
- あずき　小豆
- なす　茄子
- かき　柿

bean and fruit vegetables

こめな　米菜

- くろまい　黒米
- げんまい　玄米
- もちごめ　餅米
- はくまい　白米
- **ごはん**　炊飯
- **かゆ**　粥
- **めん**　麺

rice vegetables

あなた様のふるさとに伝わる美味な七つ菜を書き加えてみませんか

はのな 葉菜

- ちゃ 茶
- みずな 水菜
- しゅんぎく 春菊
- みつば 三葉
- ねぎ 葱
- こまつな 小松菜
- たまねぎ 玉葱
- はくさい 白菜
- ほうれん草

leaf vegetables

きのこな 茸菜

- しいたけ 椎茸
- まいたけ 舞茸
- えのきたけ 榎茸
- まつたけ 松茸
- ひらたけ 平茸
- しめじ 占地

mushroom vegetables

くきな 茎菜

- さといも 里芋
- れんこん 蓮根
- しょうが 生姜
- やまいも 山芋
- たけのこ 筍
- ほんわさび 本山葵

stem vegetables

ねのな 根菜

- ごぼう 牛蒡
- さつまいも 薩摩芋
- だいこん 大根
- にんじん 人参
- ゆりね 百合根
- かぶ 蕪

root vegetables

ふたり膳③国産七つ菜

ふるさと菜園のWASHOKU七つ菜

私たちふたりは、日本伝統の花を描き、日本伝統の野菜を研究してきました。それぞれの先祖から伝わる工芸器や茶陶器を守ってきました。幸運なことに先祖のふるさとは水と食材が豊かです。左の写真は父母が趣味で拓き、兄夫婦が守る「ふるさと菜園」で収穫されたWASHOKU七つ菜の大根。真っ直ぐ清く、旨く、わが兄らしい作品。手を加えたら、勿体ない。味を加えても、勿体ない。兄ある幸福に、感謝。

七つ菜(なな)に
手(て)をくわえたら
もったいない
味(あじ)をくわえても
もったいない

"Seven Vegetables" for *Washoku* grown in the farms of our hometown

We both have painted traditional Japanese flowers and researched traditional Japanese vegetables. We have also preserved arts and crafts as well as teaware that we inherited from our ancestors. Fortunately, our hometown where these ancestors lived is rich in water and cooking ingredients. The photo on page 75 is a Japanese radish, one of the "Seven Vegetables" for *Washoku*, that was grown on a farm first cultivated by our parents and then maintained by our brother and his wife. As a straight, clean and delicious vegetable represents well the personality of our brother, it is a shame to overcook or add seasonings to it. We are thankful for the happiness of having a wonderful brother.

ふたり膳④ ふるさと大根　Essence ④ Japanese radish from our hometown

標高1000mを越える「母なる山」が 日本には724座もあります[注]
山の栄養が里に野菜を川海に魚貝を創る…先祖から伝えて来ました
わが家のふたり膳は四季の七つ菜に感謝…次へ伝えたい暮し方です

[注] 国土地理院
日本の山岳標高
一覧から数える

ふたり膳④ ふるさと大根

ふたり膳 ⑤ 愛しき野菜　Essence ⑤ Lovely vegetables

刻まないで 砕かないで 焼かないで

この国の鳥も樹も、花も菜も、親も子も。生きものは皆、等しく菩薩さま。私たちふたりは「菩薩禅」を学んでいます。あふれる生命と仲良くして活力ある日々を貰う。これが家庭WASHOKUの恩恵です。この国の営農家が大事に育てた「七つ菜」に多く出会いました。今日収穫された「七つ菜」をわが家の台所で洗うと瑞々しい姿で必ず話しかけて来ます。私たちを、刻まないで、砕かないで、焼かないで‥ね。

きざまずに
くだかず
焼（や）かず
愛（いと）しき野菜（やさい）
ひとくちサイズ

七つ菜の「みず菜」収穫

Do not chop, crush or bake the fresh "Seven Vegetables"

Birds and trees, flowers and vegetables, and parents and children, all living things in this country are equally bodhisattvas. We are both learning the Zen of Bodhisattva. Being kind to natural life enables us to have active and vital daily lives. This is the blessing of pursuing *Washoku* at home. We encountered a number of outstanding vegetables carefully grown by professional farmers in Japan. When we wash these freshly-picked "Seven Vegetables" in the kitchen, their juicy appearance always appeals to us not to chop, crush or bake them.

わが家のふたり膳は 野菜類をきざまず ひと口大で戴きます 大切に育てられた野菜の「姿を残す・旨味を残す大きさ」に… 「野菜に金気を移さない」ようセラミックの調理具を用います

こまつ菜・みず菜

菜花のつぼみ菜

清い水で 洗う〜蒸す〜炊く〜煮る

野菜の茎や葉裏に「体毛」が無数に在る。霧や夜露を付着させて、虫や粉じんから身を守る自己防衛です。私たちは野菜を浄水で洗って、水で蒸す・炊く・煮る。「七つ菜」を水で清めて食す、がWASHOKUの真髄。穀物類の殻（から）に、豆類の莢（さや）に不味い成分をおおく含むのは、昆虫や動物から身を守る自己防衛。脱穀精米した玄米白米には、殻の微粉末が付着しているから、「米洗い初水」に飲んでおいしい水を使う。

空（そら）に大地（だいち）に水（みず）満（み）つる
この国（くに）だから
水（みず）で蒸（む）す
水（みず）で炊（た）く

ふたり膳⑥ 水で料理　Essence ⑥ Cooking only with water

Washing, steaming, boiling and simmering with pure water

There exist uncountable hairs on the stems of vegetables and on the underside of their leaves. These hairs are to protect the edible part of vegetables from insects and dirt. We wash vegetables and then steam, boil or simmer them with clean water. Purifying the "Seven Vegetables" with water is the essence of *Washoku*. The chaff of crops and the hulls of peas include plenty of bitter elements in order to protect themselves from insects and animals. As white rice that has just been threshed comes with powdered chaff, we wash rice with water that is fresh and tasty enough for the rice itself to absorb.

小菊を茹でる

小蕪を蒸す

わが家のふたり膳は「豊かな水が育てた七つ菜」を水や湯で調理（蒸す・炊く・煮る・茹でる）して体を温める野菜＝温体菜（あたたまろ〜な）で戴きます　菜は柔らかで胃腸を守り、春夏秋の梅雨冷えや冬冷えから体を守ります

ふたり膳⑥水で料理

旨み出し汁〜つぶ味噌〜万能ゆず酢

水で清めた「七つ菜」を、三つの天然調味で高めるのが、私たちふたりのWASHOKUの味わい方。天然調味一つ目は、椎茸・鰹節・真昆布の旨み出し汁。ひと口大の蕾菜・葉菜・茸菜に匙1杯かけて食べる。天然調味二つ目は、麦こうじと大豆の醸造つぶ味噌。ひと口大の茎菜・根菜に2粒〜4粒を付けて食べる。天然調味三つ目は、柚子みかんを手絞りしたゆず酢。野菜も魚肉も3滴で、味覚すっきり、香り爽やかに。

七つ菜の「里芋」をつぶ味噌で

Dashi stock with *Umami* flavor, grained miso paste, and multi-purpose Japanese citron vinegar

Boosting the flavor of the "Seven Vegetables" by purifying them with water and then adding the three kinds of pure seasonings is our own way of enjoying *Washoku*. The first natural seasoning is *Dashi* stock with *Umami* flavor extracted from Japanese mushrooms grown on natural logs as well as dried bonito shavings and fresh kelp. We enjoy bite-size sprout, leaf and mushroom vegetables with a spoonful of the stock. The second natural seasoning is fermented and grained miso made from malted wheat and soybeans. We enjoy bite-size stem and root vegetables with several grains of miso. The third natural seasoning is hand-squeezed Japanese citron vinegar. This vinegar enhances freshness in the flavor of vegetables as well as that of meat and fish.

ふたり膳⑦ 天然の三味　Essence ⑦ Three kinds of pure seasonings

旨み出し汁
つぶ味噌
ゆず酢
七つ菜に三味
合わせて添えて

わが家のふたり膳はその日の七つ菜の良さに即応する三味の試みです
始めに「七つ菜が活きる旨み出し汁」を覚え 次に「塩・砂糖・醤油に替わるつぶ味噌」を試し 次に「肉魚野菜を香り爽やか万能ゆず酢」を学びました

つぶ味噌

万能ゆず酢

旨み出し汁

ふたり膳⑦ 天然の三味

世界でjapanと呼ばれる漆器が主役

わが家を訪れた欧米アジアの方々がThis is japan!と呼んでくれる膳の主役は、桃山江戸時代の花美術を継承した、手づくり陶器の茶器類と手描き漆器類。日本の花を愛した先祖が使い、我らに伝えた器です。茶碗に米菜・実菜・茸菜の「菜入りご飯」を作り、漆器椀に根菜・茎菜・葉菜・蕾菜の「菜いっぱい汁」を作る。七つ菜で工夫した「一飯一菜」を、先祖の器で食する。二品で栄養十分。創造的WASHOKU生活は愉しい。

Lacquerware acknowledged as "japan" by the world for the central figure of our meals

The main figures of our meals to entertain foreign guests are hand-made earthen teaware and hand-painted lacquerware, all owing their style to the flower paintings of the *Momoyama* to *Edo* periods in the 16th to 19th centuries when dynamic and splendid artworks were actively produced. Our guests from the US, Europe and Asia acknowledge them as "japan." These wares were used by our ancestors who loved traditional Japanese flowers and passed them down to us. We prepare rice mixed with bean, fruit, and mushroom vegetables in an earthenware bowl, and soup full of root, stem, leaf and sprout vegetables in a lacquer bowl. "A bowl of rice with a bowl of soup" made from the "Seven Vegetables" provides enough nutrition for a day. The creative life-style of *Washoku* is pleasant.

先祖が愛でた
漆器 陶器
なおして使って
伝えますとも

ふたり膳⑧ 先祖の器 Essence⑧ Tableware from our ancestors

鳴海織部

青織部

透き漆塗

わが家のふたり膳は先人が好んだ形・色・柄の漆椀漆膳と陶器茶碗が先生です
欧米で**japan**と呼ばれる工芸漆器…ひとつでも「御かゆ・御はん・御さい・御しる・御ちゃ」を学べる万能抹茶碗…WASHOKUの器を次世代に伝えたい

ふたり膳⑧ 先祖の器

83

「咲く咲く」と歌う食用花の愛らしさ

日本の春は桜花が満ち満ち、日本の秋は菊花が満ちる。小さな花弁の桜と菊を、日本人はみんな愛しています。日本の1か所に1000年前から花咲かす写真の小菊は花の直径が4cm、筒袖のような花弁は2cm。なんとも愛らしい。私たちが「千年小菊」と呼んでいる、美味しい食用花です。花なのに噛むと野菜のような音を出す。口の中で「咲く咲く」と聞こえるから面白い。季節の花や野菜の歌声を聞こう、がWASHOKUの真髄です。

秋の膳① 千年小菊 Autumn ① Tiny Millennium Chrysanthemum

Lovely edible flowers that sing in flowery and crunchy voices

The Japanese spring is full of cherry blossoms while its autumn is filled with chrysanthemums. All Japanese love cherry blossoms and chrysanthemums with their tiny petals. Some chrysanthemums are as small as four centimeters with two-centimeter petals in a bell-sleeve shape. This is a lovely and delicious edible flower that we call the " tiny Millennium Chrysanthemum." Despite being a flower, it makes an exciting crunch in the mouth like a vegetable when chewed. It is also the essence of *Washoku* to enjoy the vigorous singing voices of flowers and vegetables in season.

この国で千年暮らす
筒袖の菊
細き花びら
いと愛おしい

秋の膳①千年小菊

古の菊 出会う

ひとつ処で
千年生きる
ちいさき菊に
学名なし
名もなき
古器の小菊
よく似た姿の
両花かがやく

花だけを摘んだ4cm小菊は料理に添える食材に仕込む

秋の膳①千年小菊

畑で摘んだ千年小菊を持帰り、畑で咲く様に活けてお供え花に　花器は 黒漆塗菓子器 金銀蒔絵江戸菊花図

筒袖ぬぎて
　さくさく
ゆず湯くぐりて
やまぶき色に
咲くさくさく
　こがね増し
筒袖ひとくち
咲くさくさく
　ああ翌秋も

ゆず酢加えた湯に３分　　　　　千年小菊のゆず酢漬け

菜花色 なのはないろ

山吹色 やまぶきいろ

秋の膳①千年小菊

2cmの筒袖花弁だけを優しくそっと引き抜く　　手絞りゆず酢の力で菜花色から山吹色に黄花の彩度が上がる

黄色い五弁花の山吹

新まい来たり
いろ淡き
米にまざりて
千年小菊

山吹膳

椀蓋とれば
五弁黄花
小菊おおきく
咲くさく

秋の膳①千年小菊

千年小菊の山吹膳／玄米入りご飯に清まし汁　お玉・湯桶・丸盆・平椀は 黒漆塗金蒔絵 松竹梅花図

「黄金の国」の輝くお米 米菜の茶碗

夏が秋に変わると、圃場は黄緑色から黄金色に輝く。脱穀精米されて黄金色に輝く玄米。半透明ガラスのように透く白米。乳白色に磨かれたもち米。黒く光る古代米。毎日2回の「米菜」料理が私たちふたりのWASHOKUの主役です。良質の米4種（玄米・白米・黒米・もち米）と、好みの陶器3種（碗・鉢・皿）で愉しむ。たわわな「みのり穂」を先祖好みの穂色陶器で迎える。

A bowl of glossy rice grown in a "golden country"

Farm fields shine gold when autumn comes. Brown rice just threshed into shining gold, white rice looking like translucent glass, sticky rice polished into milky white, and ancient rice gleaming black. Having rice dishes twice a day is the principal element of *Washoku* for us. We both enjoy the quality rice of these four kinds, namely brown, white, black and sticky rice, in our favorite earthenware. We welcome the delivery of golden harvested rice from farm fields by putting it in the earthenware of the same color with the spirit of thankfulness.

秋の膳②みのり穂　Autumn② Harvested rice

みのり穂
育てた
まもり水
田にむかえゆく
秋日幸あり

たわわな「みのり穂」と、育てた水源の湧水「守り水」をいただき、黄金色した茶陶器の「水田」に活けてお供え

秋の膳②みのり穂

黄瀬戸で迎える

昔のひとも
お米のうつわ
みのり穂に似た

いま新米供えし
桃山うつしの
　穂いろ茶碗

白　米
はくまい

玄　米
げんまい

守り水に活けたみのり穂を束ねた頃に
圃場で刈入れ収穫した新米が2つ届き
黄瀬戸のご飯茶碗に入れて御供えする
（左）玄米　（右）精米した白米

秋の膳②みのり穂

今朝の菜
みのり穂
さつまいも
かゆとなり
相性よき
いんげんサラダ
添えにけり

今宵の菜
みのり穂
ぶなしめじ
ご飯となり
相性よき
白菜にんじん
添えにけり

みのり穂
みぶなしらす
菜飯となり

あーと菜

相性よき
花がた長いも
添えにけり

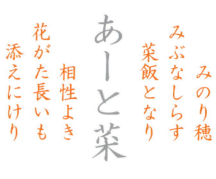

みのり穂
もち米あずき
赤飯となり

ふたり膳

相性よき
大根こまつ菜
添えにけり

秋の膳②みのり穂

この国の大地・草木・
花・実・鳥を想わせる
陶器を日々愛用する

多し菜 少し菜 その日好み

なるみ織部
え唐津
ねず志野
桃山うつし

大地の器
あたたか菜
花絵の器
すずしい菜

米とおなじ湧き水で育った「幸福な豆」

標高1000m余の山から導く湧き水を同じ水源で使う日本の営農家さま。一斉に米作3年、一斉に豆作3年。同じ圃場で3年ごと交互に作付してほしい。農薬を使わないで「幸福な米」と「幸福な豆」になる期待からです。米豆は互いのタンパク成分を高め合うベストぱーと菜。米と豆をいっしょに食べよう、がWASHOKUの真髄。湧水美田で育った「幸福な豆」からピュア豆乳を作って旨し、「玄米めん白米めん・自家豆乳つけ麺」でまた旨し。

夏はちいさき　紫の花
秋はおおきな　二つ実に
ふうふ福豆

Happy beans grown with pure spring water the same as rice

Japanese farmers in the same community share pure water from a mountain about 1000 meters high. All these farmers simultaneously grow rice and beans every three years in turn with no or little pesticide. They do so, being motivated by their strong wish to grow "Fortunate Rice and Beans." Eating rice and beans together is the essence of *Washoku*, because rice and beans are ideal partners that mutually enhance their proteins when eaten together. We enjoy making soymilk from "Fortunate Beans" grown with pure spring water, and also using it as soup for noodles made from brown and white rice.

秋の膳③ ふうふ福豆　Autumn ③ Couple beans

春収穫する「空豆」は花弁3.5cmの白花が春に咲き、秋収穫する「大豆」は花弁1.5cmの可憐な紫花が真夏に咲く

秋の膳③ ふうふ福豆

お米のように
滋養のもと
1000m余の
山から導く

湧き水そだち ふうふ福豆

清水にひたせば
ふっくら二倍
朝な夕な
お米のように

右の丸い二つ実福豆を水に1日
ひたすと…左の様に2倍に戻る

自家豆乳つけ麺

お米と福豆
ベストぱーと菜
互いのたんぱく
　高めあう
豆乳ぞうすい
豆乳おかゆ
滋養は倍なり
　長寿の膳

湧水美畑で育った大豆80gを使って「おからとピュア豆乳」を2人分つくった…さらりと濃い！

米めんを 手づくり豆乳旨み出し汁で「玄米めん白米めん・自家豆乳つけ麺」 器は 鼠志野脚付角鉢と鼠志野深向付

秋の膳③ ふうふ福豆

秋の膳④ 黒豆舟遊び　Autumn ④ Boats of black beans

月花の蒔絵漆器に「豆の舟」を浮かべて

9月中旬の満月を迎えると、日本の家庭では「観月」の膳を縁側や窓際に設ける。1000年前の京都貴族の邸宅では庭園の池に船を浮かべて、池の水に映る月を愉しむ、杯に映る月を楽しむ「舟遊び」が流行ったという。旬の食材で季節の風情を演出するのがWASHOKUの真髄。私たちは黒大豆枝豆の収穫期10月に「観月」の膳を設けて、茶入れと菓子の小さな盆黒漆の池に豆さやの緑舟を浮かべる。早摘み黒豆は甘く柔らかい。

ふたつ実の早摘み黒豆は
二槽並んで舟遊び

秋（あき）の夜（よ）は
月花（げっか）の蒔絵（まきえ）で
舟遊（ふなあそ）び
ひとつふたつ実（み）
黒豆（くろまめ）が舟（ふね）

Floating bean pods as tiny green boats in lacquerware with the pattern of a flower called "queen of the night"

At the full moon in mid-September, Japanese families prepare meals for the full moon on a veranda-like porch at home. Aristocratic families in *Kyoto* 1000 years ago enjoyed the reflection of the moon onto ponds in their residences where they were boating and onto the surface of small sake cups on the boats. Describing the beauty of seasons with seasonal ingredients is the essence of *Washoku*. We prepare meals for the full moon in the autumn when black beans become seasonal by, for example, putting bean pods onto a lacquer tray as if they were tiny green boats floating on a pond. Freshly picked black beans are soft and sweet.

ひとつ実の早摘み黒豆は 平棗の上で「舟遊び」 平棗は直径8.5cm 黒漆塗金銀蒔絵 月と須々木図

秋の膳④ 黒豆舟遊び

- 淡萌黄 うすもえぎ
- 桔梗色 ききょういろ

早摘み黒豆は 大きく甘く柔らかい　豆サヤは淡萌黄で 実の皮がうす紫色や濃い紫色

黒まめ 桔梗(ききょう)染め

うす萌黄
早摘み
豆莢ぬぎすて
舟遊び

池の花いろ
ちかづきて
みな染まりし
青むらさき

半月盆の桔梗の池で舟遊び　黒漆塗金銀蒔絵 桔梗図

秋の膳④ 黒豆舟遊び

黄瀬戸の茶碗に「早摘み黒豆」お月見ごはん

黒まめ月も月見ごろ

満月に
あわせて
お月見
ああ美し

収穫に
あわせる
お月見
ああ旨し

月映す蒔絵皿の池で黒豆舟遊び
（左）月と須々木図、（右）月と萩図

秋の膳④ 黒豆舟遊び

秋の七草と食器で「一菜一絵」を愉しむ

一期一会を、一菜一絵で。これこそWASHOKUの真髄でしょうか。故郷の兄や営農家から年にいちど届く季節野菜で「一菜」を作り、私的食器類で「一絵」を作る。思い出に残るオリジナル膳を愉しんでいます。ことしの秋は、七草のひとつ＝本葛の粉で「本くず湯」を作り、江戸くず花の蒔絵膳でデザートらしく演出。さらに「本くず栗つつみ」を作り、江戸花の秋菓子膳も。

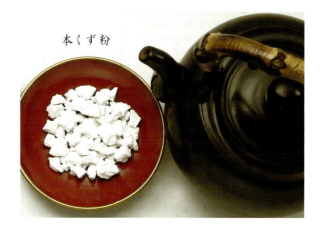

本くず粉

いにしえの
きんぎん蒔絵（まきえ）
むき合（あ）いて
本（ほん）くず愛（め）でし
わけを知（し）る

"One vegetable, one image," presented by the seven autumn herbs and tableware

"One time, one meeting" is a Japanese four-character idiom, meaning that we should treasure every encounter for it will never recur. Representing this spirit with "One vegetable, one image" is the essence of *Washoku*. We cook one of the vegetables delivered only once a year from brothers and farmers in our hometown and then creatively arrange it on artistic tableware. Last autumn, we made hot water with *Kuzu*, one of the seven autumn herbs, and created a dessert plate for autumn by arranging a cup that contains this hot *Kuzu* water on a lacquer tray patterned with *Kuzu* flowers. We also made a *Kuzu* jelly cake with a whole chestnut in its center, and enjoyed a sweets plate for autumn by arranging it on a lacquer tray with silver and gold flower reliefs.

秋の膳⑤ 本くずの絵　Autumn ⑤ Painting of Kuzu

柚子ジャムの「本くず湯」をお供え　脚付膳は 黒漆塗金銀蒔絵 江戸くず花図

秋の膳⑤ 本くずの絵

柚子ジャムの本くず湯
器は 尾形乾山写し 秋の花図

おみなえし
　はぎ
　　ききょう
　　　すすき
　　　なでしこ
　　　ふじばかま
　秋暮れどきの
　　本くず湯

秋の膳⑤ 本くずの絵

本くず栗つつみ

秋の実菜
優しくつつむ
よき菓あり
実り伝える
色足さず
細工せずとも
蒔絵に負けぬ
よき菓なり

栗あん30g＋本くず20g＋栗甘露煮6個＝ふたり分

ふたり膳の秋菓子「本くず栗つつみ」　菓子鉢は 黒漆塗金銀蒔絵 江戸菊花図　菓子皿は 黒漆塗金銀蒔絵 花椿と桔梗

秋の膳⑤ 本くずの絵

抱一さまの 葛(くず)すすき

夏すぎ秋に
なりぬれば
この絵に尽きる

食材活かす
江戸絵師の技
いま活けてみる

秋の膳⑤本くずの絵

葛はマメ科
紫色の花が
豆莢になる

右頁の絵は 江戸の画家・酒井抱一「夏秋草図屛風」部分図　本葛の蔓は細いが力強く空中に数メートル伸びて行く

真白い芸術的根菜で「冬の花」咲かす

五穀に次ぐ理想的な野菜をめざし、千年かけて完成した根菜が「小蕪」です。まるく太った根は甘く、葉も茎も柔らかい。一年通して収穫できるが、盆地や谷間の霧が育てる冬の小蕪は芸術的。真白き肌に凛とした立ち姿が美しく愛おしい。食材を余さず全部食べるのがWASHOKUの真髄。この国が育てた美人野菜を「霧蒸し」して、白い花のように咲かせます。

冬の膳① 霧の小蕪　Winter① Tiny turnip grown in mist

Snow-white turnip blooming like winter flowers

The tiny turnip was brought to perfection over 1000 years, producing the most ideal vegetable next to grains. The plump root of the tiny turnip is sweet, while its leaves and stems are soft. A winter tiny turnip grown in mist in a basin or valley is esthetic with its white skin and dignified shape. Eating an entire ingredient without any wastes is the essence of *Washoku*. We cut and open a tiny turnip in the center and then steam it in mist, so that it can look like a white flower.

陽(ひ)やわらぐ
薄霧(うすきり)に守(まも)られて
萌黄(もえぎ)も白(しろ)も
まぶしい小蕪(こかぶ)

冬の膳①霧の小蕪

小蕪は芸術(アート)なり

五穀の次に
国野菜となれ
1000年を
越えて生まれし

ましろき肌に
腰高すがた
根やわらかく
葉茎ぜんぶ旨し

冬の膳① 霧の小蕪

小蕪のアート(1) 凛とした立ち姿に敬意をはらい、裃をまとったようなお供えに

小蕪のアート(2) 古九谷25cm絵皿の上で 根は白い花に 茎は緑の筏舟に 花芯に「つぶ味噌」そえて

山の霧が育てた
お野菜だもの
霧蒸し8分
家で花さいた

霧蒸し小蕪　冬の花

冬菜みな
きれい好き菜
蒸し好き菜
国野菜さかそ

冬は体に優しい「低カロリー」のおやつ

秋の紅葉が終ると、山は冬になる。一方で里の畑は春夏秋に蓄えた栄養で、五つ菜（根・茎・葉・蕾・実）を冬に次々と生産する。五つ菜の主役は栄養があって体に優しい「里芋」。根菜でなく茎が大きく育つ茎菜。芋類の中で「低カロリー」で、おやつ・夜食向きです。私たちふたりは、里芋をしっかり蒸して二つに割り「花いかだ」の舟にして、萩咲く漆膳の池に浮かべる。

花いかだは葉の中心で花が咲き実が育つ

冬の夜は からだ温まる 里の芋 萩さく池に 花いかだ

Nutritious yet low-calorie snacks in winter

Japanese mountains say hello to winter when the red and yellow leaves are gone. On the other hand, five kinds of vegetables, namely root, stem, leaf, sprout, and bean and fruit vegetables, grow one after another in this season in farm fields at the bottom of these mountains with nutrients stored from spring to autumn. Rich in nutrients and mild in flavor, *Sato* potato is a key vegetable among these. Moreover, being a stem vegetable, not a root one, and relatively low-calorie among the range of potato varieties, *Sato* potato is good as a snack for day and night times. We steam and cut a *Sato* potato into two pieces, and then arrange them in a lacquer tray patterned with Japanese bush clover as if they were rafts in a pond surrounded by these flowers.

冬の膳② 里芋の花筏　Winter ② Rafts of Sato potato

低カロリーくき菜「里芋」の霧蒸し花いかだ膳　角盆は 黒漆塗金銀蒔絵 萩花図

冬の膳② 里芋の花筏

花筏は 葉の中心でつぶ花が咲く

里芋は根ではなく 茎がおおきくなった「茎菜」
芋類の中では低カロリーなので間食・夜食向き
霧蒸しすると そのまま食卓に映える容姿端麗

低カロリーくき菜

蒸したさといも
　人という字に
　　なりにけり
　人にやさしい

親いもひとつに
　子いも孫いも
　かぞくおもいも
　　いちばん嬉し

冬の膳② 里芋の花筏

冬の膳③ 万能ゆず酢 Winter ③ Multi-purpose Japanese citron vinegar

果実の酢は 人と菜の「水を清く」する

食はさっぱり、さわやかに（清い状態）。器はすっきり、すずやかに（涼しい状態）。清涼な食感と盛付の両立がWASHOKUの真髄。野菜の中身も人間の体も70％以上が水。私たちの調味料は、人体と食材の水を清くする「ゆず酢」が主役です。ゆずの中味を搾って作る天然酢は少しの量ですべての野菜から魚肉をさっぱりおいしくする。ゆずの外皮も写真のようにきざんで、摺って、ジャムにして。ぜんぶ使います。

本柚子
手しぼる
ゆず酢あり
黄皮に香りに
ちから有り

柚子は皮を下にしてぎゅっと絞り
黄皮の油胞＝香り成分を取り込む

Purifying water running in both vegetables and human bodies with fruit vinegar

Food should be fresh and pure, while tableware should be simple and cool. It is the essence of *Washoku* to realize both the fresh flavor and the cool arrangement of ingredients at the same time. Water makes up more than 70% of vegetables and human bodies. Our principal seasoning is Japanese citron vinegar that purifies the water contained in vegetables and human bodies. Natural vinegar made from the juice of the Japanese citron adds fresh flavor to all vegetables, fish and meat, even if only a spoonful is used. We use the entire Japanese citron, not only by making vinegar from its juice but also by chopping and grating its peel.

冬の膳③万能ゆず酢

天然実生の柚子で作った「万能ゆず酢」は純銀梅花文様デカンタに…上部の厚い「油脂膜」が香りと味を守り保つ

柚子皮は丸く削いで
汁椀や紅茶に浮かし
香りを愉しむ

柚子皮は細く切って
煮込みジャムを作り
甘味を愉しむ

夏の柚子玉は
半分に切って
酸味を愉しむ

柚子ジャムを
通年愉しむ

柚子皮は摺りお
ろしトッピング
で香りを愉しむ

夏秋 冬春 ゆず 愛し

かおり
酸味
夏の一寸
緑ゆず玉
冬の三寸
黄ゆず玉
美膳に
万能

冬の膳③ 万能ゆず酢

冬の膳④ 立花ねぎ　Winter ④ Leeks in the Rikka style

冬の畑に「在る」が姿に活けて感謝して

自宅の居間に季節の花木を真っすぐ立て、自然景観を表現する伝統的な生け花が600年前から日本にある。「立花」と書いてRIKKAと読む。12月に営農家夫妻から贈られる長身90cmのねぎは、美しく愛しいから立花にします。畑に「在る」姿に、「お二人で舞う」形に。食べる前に感謝を捧げるのがWASHOKUの真髄。長身ねぎは小松菜や白菜と同類で、全身が葉っぱの菜。冬のねぎは真ん中に「ムチン」を多く蓄え、これが旨い。

Arranging vegetables in a vase as they are in farmland with the spirit of thankfulness

A traditional way of flower arrangement called *Rikka*, arranging fresh plants and flowers in a vase to represent their natural beauty at home, was developed in Japan 600 years ago. We arrange Japanese leeks as long as one meter sent to us by a farmer couple every December, in a vase at home as if they were in farmland to portray their beauty and loveliness. It is the essence of *Washoku* to express our thankfulness for ingredients before eating them. Though long, the Japanese leek is one of the leaf vegetables, like spicy spinach or nappa cabbage. It becomes even more delicious during winter, containing rich mucin, heavily glycosylated protein, in its center.

ふたりして
舞(ま)うがごとき
立花(りっか)ねぎ
この姿(すがた)みて
ことしも暮(く)れる

冬の膳④ 立花ねぎ

ぜんぶ旨しねぎは葉菜(はのな)

　くきは
　　足元一寸
　一尺しろ葉
　二尺みどり葉
白は花に緑は鳥
今年の暮れは
　立花ねぎの
　　花鳥蒸し

立花ねぎは 小松菜・白菜・玉葱と同じぜんぶが葉菜 白葉で五弁の花を 緑葉から千鳥をつくって霧蒸しする

師走(しわす)に嬉しとろとろムチン

寒さと湿気
初雪の畑
青葉いならぶ
立花ねぎ

しろ葉と
みどり葉
立花の真なか
ムチン増す

立花ねぎ
青葉色

ムチンの成分は「たんぱく質と糖質」、滋養強壮効果や冬風邪から喉や胃を守ります

とろとろムチンゆず塩たれ

冬の膳④ 立花ねぎ

緑米

赤米

黒米

もち米

玄米

白米

冬の膳⑤ お米おせち

140

先祖を迎える
　ふたり膳
みなさま好みの
　お米おせち
　ご一緒に

黒漆塗金蒔絵 四季花図の
お重で正月を迎える

お正月膳はWASHOKUの全てが詰まる

主役として米料理を食べる、がWASHOKUの真髄。新年のわが家では、「お米おせち」と名付けた3膳です。一の膳は、白米もち・小松菜椀、玄米もち・水菜椀、ひとりでふたつの椀。二の膳は、玄米・白米・黒米の祝い飯、白米・玄米・緑米の祝い飯。漆蒔絵重箱の真ん中に紅白ごはん。四隅には4つの野菜料理でお弁当風にします。三の膳は、お抹茶に祝い菓子。もち米落雁粉で手作りした「松と椿」。WASHOKUのすべてを詰めました。

新ねんもち菜椀ふたつ

玄米もちと
みず菜
白米もちと
こまつ菜
このみの花椀
好みの七菜
お米おせちの
一の膳

冬の膳⑤ お米おせち Winter ⑤ Rice Osechi

Packing everything about *Washoku* into *Osechi*, a New Year's meal box

It is the essence of *Washoku* to eat rice as the principal dish of a meal. We celebrate the new year with a three-layered meal box filled with rice dishes, called "Rice *Osechi*." The box on the top has two bowls: a white rice cake with Japanese mustard spinach as garnish, and a brown rice cake with Japanese greens called "water greens" in Japanese. The box in the middle is filled with two flower-shaped rice balls: one is made with brown, white and black rice, and the other is made with brown, white and green rice. Four vegetable dishes are also arranged in each corner of the box. The box at the bottom consists of sweets using powdered sticky rice and white sugar and then molded into the shapes of pine trees and camellia flowers. They come with bowls of green tea. We pack everything about *Washoku* into this "Rice *Osechi*."

もち菜椀の2＝玄米もちと
たっぷり水菜
椀は 漆塗金銀赤絵 菊花づくし

もち菜椀の1＝白米もちと
たっぷり小松菜
椀は 漆塗金銀赤絵 菊花づくし

冬の膳⑤ お米おせち

二の膳は
四季花図を描いた
黒漆塗金蒔絵お重

年はじめ 紅白ごはん

玄米白米
黒米の祝い飯
白米玄米
緑米の祝い飯

好みの古代米
まんなかに
お米おせちの
二の膳

冬の膳⑤ お米おせち

もち米 祝い菓子

年越しの鐘
らくがん作り
あさひ松
ゆきつばき

好みの古陶
初茶たて
お米おせちの
三の膳

口に入れるとふわっと溶ける「もち米落雁粉」の祝い菓子を作る
右は「ゆきつばき」左は「あさひ松」初茶は赤楽茶碗と黒楽茶碗で

146

もち米お供え

冬の膳⑤ お米おせち

雪が降った日に 春を呼ぶ「蕾菜」かゆ

1月には珍しく雪が降り、テラスに積もる。寒い日に「からし菜のつぼみ菜」と「菜花のつぼみ菜」が届いた。日本名でMOEGI、英語ではSpring Greenという。萌黄とは春の成長を知らせる生き生きした緑色です。ふたつの蕾菜を私は霧蒸しする。雪笹陶器に盛った粥に、たっぷり沈めて命名する。春を呼ぶ蕾菜かゆ。国産野菜で季節の訪れ知る、がWASHOKUの真髄。

淡萌黄 うすもえぎ

萌黄 もえぎ

Rice porridge with sprout vegetables to call for spring on a snowy day

Last January, we had snow falling on our terrace for the first time in many years. On such a cold day, the sprout vegetables "mustard spinach" and "field mustard" were delivered to our home. They were *Moegi*-colored, a color which means spring green in Japanese and speaks of the development of spring. We steamed these two sprout vegetables with mist and soaked plenty of them into rice porridge in an earthenware bowl with a pattern of snow and bamboo, naming the dish "rice porridge with sprout vegetables to call for spring." It is the essence of *Washoku* to learn the arrival of seasons through vegetables produced in Japan.

冬の膳⑥ 蕾菜かゆ　Winter ⑥ Sprout vegetables

冬の膳⑥ 蕾菜かゆ

冬のつぼみ菜
霧蒸して
器いっぱい
萌黄いろ
眩しかりけり

春よぶ蕾菜かゆ

菜花の
　つぼみな
からし菜の
　つぼみな
萌黄芽吹く
雪笹の器
からだ温まる
雪日たのし

からし菜のつぼみな

菜花のつぼみな

霧蒸し蕾菜たっぷり「春よぶ蕾菜かゆ」 大鉢と向付は 尾形乾山写し雪笹図

冬の膳⑥蕾菜かゆ

野に咲く花 野菜の花 みんな愛しく

青い山に、白い雪が残る。冬が終る頃、池の端、川の淵、盆地の畑に咲く黄の花がアブラナ科「菜花」です。「小蕪」「こまつ菜」「白菜」「ちんげん菜」―冬の美味な菜はアブラナ科、よく似た四弁の黄花を咲かせます。野に咲く花々が好きで、畑の花々も可愛い。その想いを短い詩歌に、接写レンズで写真にします。生きものを在るがまま愛する心がWASHOKUの真髄。

冬の膳⑦ 四弁の黄花 Winter ⑦ Yellow flowers with four petals

Loving both flowers in the field and the flowers of vegetables

When white snow still remains on green mountains towards the end of winter, cruciferous leaf vegetables bloom with yellow flowers in farms that are located by ponds, rivers and basins. All delicious winter vegetables, such as "tiny turnip," "mustard spinach," and "nappa cabbage," are cruciferous vegetables that bloom with similar four-petalled flowers. We translate our affection for flowers in fields and farms into short poems and photos. It is the essence of *Washoku* to love living nature as it is.

菜花(なのはな)こまつ菜(な)
よく似(に)た四弁(よひら)の
黄花(きはな)さく
ああもうすぐ春(はる)

冬の膳⑦ 四弁の黄花

小蕪の四弁黄花

ちんげん菜の四弁黄花

菜花の四弁黄花

こまつ菜の四弁黄花
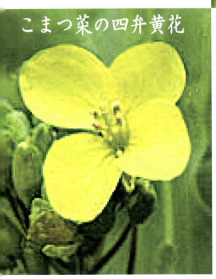

はく菜の四弁黄花

四弁のごちそう

下ごしらえ　あって柔らか
下ごしらえ　無くて爽やか
春夏かかさず　黄花の菜
秋冬きらさず　四弁の菜

菜花色　なのはないろ
淡萌黄　うすもえぎ
萌黄　もえぎ

冬の膳⑦　四弁の黄花

原 伊路波「天空の花画シリーズ」金雲山桜菩薩図

85cm × 32cm

金雲(きんうん)そだち
御花(おはな)さま
若葉(わかば)めじろ
花(はな)つぼみ
芽(め)ぶきの合唱(がっしょう)

私たちふたりは
詩歌で伝える
WASHOKU七つ菜
を表現研究しながら
桃山江戸の花美術
を創作しています

原 伊路波「天空の花画シリーズ」青雲蓮華菩薩図

85cm × 32cm

青雲（せいうん）そだち
しろい花（はな）つむ
蓮華（れんげ）さま
ふたりで描（えが）く
祈（いの）りいろ

伝統の花々には
菩薩さまが居られる
と互いに感じて
花画で伝える
BOSATSU七つ花
の連作になりました

日本の花をふたりで描く Painting of Japanese flowers together

花画で伝えるBOSATSU七つ花

先祖が愛した「桃山・江戸・明治の花美術」を私たちも大切にし、和食研究と共に花画の研究をしています。光琳さま・若冲さま・抱一さまたち江戸絵師が愛でた花々を描きたくて、木版画や植物図譜に残され今に伝わる花を根気よく探す。金雲天空に「山桜」を描き青雲天空に「蓮の花」を描く。二つの花に別々の菩薩さまが現れて…感じたまま描き入れ…連作になる。花画で伝えるBOSATSU七つ花、一度ごらんあれ。

Practicing the Heart of Bodhisattva with the "Seven Flowers"

We attach affection to flower paintings in the *Momoyama*, *Edo* and *Meiji* periods that our ancestors also favored, and study these flower paintings in addition to *Washoku*. Hoping to paint the same great flowers that the famous painters of the *Edo* period, such as *Ogata Kōrin* and *Sakai Hōitsu* of the *Rimpa* school as well as *Itō Jakuchū*, used as motifs in their woodprints and botanical arts, we patiently search for these flowers all over Japan. We draw wild cherry blossoms in the sky with golden clouds and lotus flowers in the sky with blue clouds on canvas. We feel the presence of different Bodhisattvas in different Japanese flowers, and include these Bodhisattvas in our paintings as we feel, making these paintings a series of the "Seven Flowers with Bodhisattva."

きみの心(こころ)に
花(はな)は咲(さ)く
いちばん
好(す)きな
はなが咲(さ)く

接写の詩

はなつきゆき
この国の四季は
遠くにあって
　美しい
先人の歌は
望遠レンズ
すこし離れて
観ながら書いた

蕾菜 茎菜 実菜
この国の野菜は
近くにあって
　愛おしい
ふたりの詩は
接写レンズ
よって寄って
視つめて書いた

山並の詩

四方の山辺
春は山桜
山里の秋
山路に雪
先人の歌は
季節の情景
山並のシーンを
好んで書いた

春みかん
1番摘み茶
みのり穂
霧の小蕪
ふたりの詩は
季節の香り
山並フォルムに
自由に書いた

感謝の詩

手描きの漆器
手づくり陶器
　先人さま
　ありがとう
日本の美と姿
守り伝えし
　先祖さま
　ありがとう

禅のこころ
食の心得
　父と母
　ありがとう
営農家の皆さま
　ありがとう
愛しき七つ菜
　ありがとう

あとがきの詩〜この国に感謝

田畑から
わが家に
くると
背をのばす
四季の野菜
カメラを
寄せると
ポーズする
「七つ菜」が
主人公の
和食の写真

茶も豆も蕪もネギも 葉を凛と立てる。
著者のカメラ前で、四季の野菜たちが
ポーズする。こんな写真が撮れるのは
著者が「七つ菜」と交遊暦六十年だから。
原伊路波は美景美食な伊豆市の生まれ。
父の影響で幼少から禅寺と山渓歩き暦
五十年、花鳥や「七つ菜」を接写する愉
しさ覚え絵画と写真が上達したという。
武蔵野美術大学を卒業し広告の電通本
社に入る。農業・食品・家電・百貨店・教
育事業等のアート・デザイナー暦四十年。

わが家に
やってくる
小さな鳥
かわいい花
お庭の樹
そして
野菜たち
父母の如く
話しかけ
いつくしむ
詩歌の本

日本とフランスは五百の伝統色がある。
共著はらきみこは家に伝わる着物帯で
人形を創作、桃山江戸の仏や花を描き、
七つ菜で膳をしつらえ、国際交流する。
『槇の木おとうさま 行ってまいります』
玄関の百年緑樹にきみこは声をかける。
『どん子 メジロ また来たね』毎年春に
来る茸菜や小鳥に伊路波は声をかける。
「先人の母性・父性の幹と枝に命満ちる
大樹のごとき家」が原夫妻の理想とか。
命の個性と色を詩歌と写真で伝える本。

この国の
「和色」の器
山の栄養
きれいな水
天然の味
「七つ菜」の
チカラで
健康子育て
健康長寿を
めざす
家庭膳の書

平安・鎌倉・室町・桃山・江戸期の千年で
「和色」の美と膳と「七つ菜」が完成した。
米菜・実菜・茸菜・蕾菜・葉菜・茎菜・根菜
「七つ菜」を平成のいま栽培販売される
100万戸の営農家を応援する本です。
「和色」で陶器・漆器や木器・鋳器を作る
愛用する300万人を応援する本です。
「七つ菜」と「和色の器」を世界に発信し
展開する100万人を応援する本です。
「子育て」「長寿」を七つ菜主体の家庭膳
で試みる世界中のあなたを応援する本。

Promoting the Essence of
WASHOKU
with Seven Vegetables

詩歌で伝える和食七つ菜®
2015年8月21日発行　2016年10月31日2刷

著　者：　原　伊路波（詩歌・撮影・テキスト）
共　著：　はら きみこ（詩歌・膳しつらえ）
翻　訳：　Kotono HARA（テキスト監修）
編　集：　いろは色波動デザイン研究所®
　　　　　表現教育研究所デザイン室
後　援：　学校法人 滋慶学園グループ
発行人：　田仲　豊徳
発行所：　株式会社 滋慶出版／つちや書店
〒100-0014 東京都千代田区永田町2-4-11-4F
Tel.03(6205)7865／Fax.03(3593)2088
http://www.tuchiyago.co.jp/

印刷・製本：　株式会社 暁印刷
※乱丁・落丁は、お取り替えいたします。

© Iroha HARA 2015 + Design Institute of Color Wave®
+ Institute of Expression & Education, Printed in Japan